Biological Terrorism

Steve Goodwin
University of Massachusetts

Randall W. Phillis
University of Massachusetts

San Francisco   Boston   New York
Cape Town   Hong Kong   London   Madrid   Mexico City
Montreal   Munich   Paris   Singapore   Sydney   Tokyo   Toronto

Acquisitions Editor: Michele Sordi
Assistant Editor: Michael J. McArdle
Marketing Manager: Josh Frost
Manufacturing Buyer: Vivian McDougal
Designer: Detta Penna

Cover image: Alex Wong/Getty Images.

ISBN 0-8053-4868-9

Copyright © 2003 Pearson Education, Inc., publishing as Benjamin Cummings, 1301 Sansome St., San Francisco, CA 94111. All rights reserved. Manufactured in the United States of America. This publication is protected by Copyright and permission should be obtained from the publisher prior to any prohibited reproduction, storage in a retrieval system, or transmission in any form or by any means, electronic, mechanical, photocopying, recording, or likewise. To obtain permission(s) to use material from this work, please submit a written request to Pearson Education, Inc., Permissions Department, 1900 E. Lake Ave., Glenview, IL 60025. For information regarding permissions, call 847/486/2635.

Many of the designations used by manufacturers and sellers to distinguish their products are claimed as trademarks. Where those designations appear in this book, and the publisher was aware of a trademark claim, the designations have been printed in initial caps or all caps.

3 4 5 6 7 8 9 10 —PBT—07 06 05 04 03

www.aw.com/bc

Contents

What Are Biological Weapons   1
Historical Use of Biological Weapons   1

Potential Biological Weapons   5

The Virus that Causes Smallpox   7
Vaccines   9
Global Eradication   11

The Bacterium that Causes Anthrax   13
Pathogenesis of B. Anthracis   15
Antibiotics and Vaccines   17

The Toxin of *Clostridium Bolulinum*   18

The Epidemiology of Biological Weapons   21

Resources about Biological Weapons   24
Especially for Students   24
Resources for Educators   24
Review Articles   25
Popular Books   25

What Are Biological Weapons?

Biological weapons should be unthinkable. They are nothing less than the intentional use of organisms to harm human beings, for the most part by causing disease. Disease is an abnormal condition of an organism, especially with regard to its physiology. The original meaning of the word is "not at ease." To understand biological weapons, we will need to understand what causes disease. We will also need to understand our own body's defenses against disease. Finally we will need to understand not only how disease affects individuals but how it interacts with large populations of individuals. Disease may be the result of inherent weaknesses of an organism, the result of environment stresses, or may be caused by other organisms. Organisms that cause disease are referred to as **pathogens** and are very often (but not always) **microorganisms**. Pathogenic microorganisms can cause diseases of humans and other animals, they can cause diseases of plants, and they can even cause diseases of other microorganisms. But before we explore the basic biology behind biological weapons, let's take a brief look at the history of biological weapon use.

HISTORICAL USE OF BIOLOGICAL WEAPONS

On October 4, 2001, a 63-year-old Florida man was reported by the Florida State Department of Health to the Centers for Disease Control and Prevention (CDC) in Atlanta, Georgia as having a case of the disease **anthrax**. Despite aggressive medical treatment, he subsequently died of inhalation anthrax. The bacterium that causes anthrax, *Bacillus anthracis*, was found both in his workplace and in the nasal passage of a coworker. What was especially unusual about this report is that in the United States there had only been 18 cases of inhalational anthrax reported between 1900 and 1978 and most of these involved people whose occupations were likely to expose them to *B. anthracis*. This was not the case with the Florida man whose occupation would not have been expected to carry any specific risk of exposure. No case of inhalational anthrax had been reported in the United States since 1978 prior to the Florida case.

Over the course of the next month several more cases of anthrax were reported. These involved workers at media companies, hospitals, postal facili-

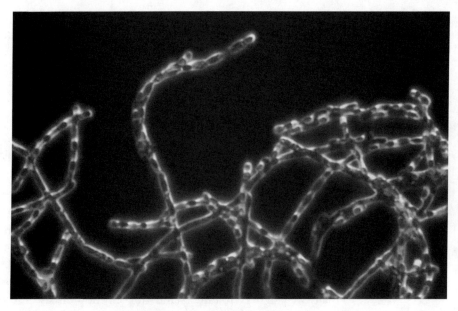

FIGURE 1. Dark field *Bacillus anthracis* with spores, 600x. Source: Michael Abbey/Photo Researchers, Inc.

ties, and the U.S. Government. It was learned that letters contaminated with the spores of *B. anthracis* were sent both to media companies in New York City and to members of the United States Congress. Spores are a dormant form of the bacterium but are able to cause the disease under the right conditions. In its November 2, 2001 issue of *Morbidity and Mortality Weekly Report* (Vol. 50, No. 43), the CDC reported a total of 21 cases of anthrax (16 confirmed and 5 suspected). As of this writing, the source of the contaminated letters has not be found but all available evidence suggests that they represent an act of bioterrorism.

This was not, however, the first suspected use of biological weapons. It was not even the first suspected use of *B. anthracis* as a biological weapon. During World War II, the British exposed sheep to *B. anthracis* on Gruinard Island off the coast of Scotland. Explosives were used to expose sheep to spores of the bacterium. Because the spores are able to survive in the environment for long periods of time, the island remained uninhabitable for 36 years. Decontaminating the island took nearly eight years and required 280 tons of formaldehyde and 2000 tons of seawater.

An accidental release of *B. anthracis* spores occurred in 1979 at a microbio-

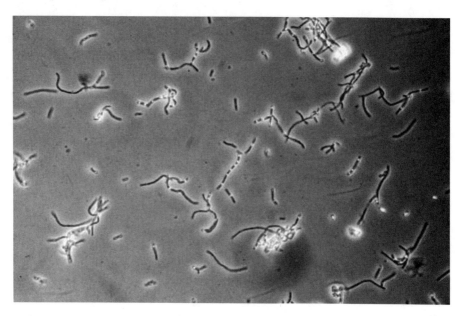

FIGURE 2. A phase-contrast photomicrograph of *B. cereus*, a close relative of *B. anthracis* which is the causative agent of anthrax. The spores appear as phase bright circles spaced along the length of the rod-shaped cells.

logical facility in Sverdlovsk in the former Soviet Union. This was a plant that was manufacturing *B. anthracis* as a biological weapon. A filter was inadvertently left off an air handler for several hours and spores were released into the surrounding countryside, resulting in 79 cases of anthrax and 68 deaths. Some of the cases developed as late as 43 days after the initial release. It has been speculated that because the local communist party boss ordered trees and buildings to be washed down after the accident, additional people may have been exposed to the spores. The local communist party boss at the time was Boris Yeltsin. At least one deliberate release of *B. anthracis* spores has been documented. In 1993 the terrorist cult Aum Shinrikyo released *B. anthracis* spores from the roof of a building in Kameido, Japan. No one was made ill by this release, a fact that demonstrates that using microorganisms as biological weapons is not necessarily an easy thing to do. This same cult is believed to have unsuccessfully attempted to release botulinum toxin in Tokyo, Japan.

Anthrax is not a recent disease. It is thought that it may have been responsible for the fifth Egyptian plague described in the Bible. A plague is a calamity or affliction, originally thought to be of divine origin. A plague that involves

TABLE 1: Potential Biological Weapons

Disease	Pathogen or toxin	Incubation period	Comments
anthrax	bacterium: *Bacillus anthracis*	up to 60 days	has received more attention as a biological weapon than any other disease
botulism	toxin produced by the bacterium: *Clostridium botulinum*	2 hours to several days	ingestion of foods contaminated with botulinum toxins causes muscle paralysis
brucellosis	bacterium: *Brucella melitensis*	3–4 weeks	a disease primarily of livestock that can be life threatening in humans
cholera	bacterium: *Vibrio cholera*	12 hours to 5 days	the toxin is actually coded for by viral DNA that is contained in the bacterium
Congo-Crimean hemorrhagic fever	virus	1 to 9 days	transmitted by ticks
Ebola hemorrhagic fever	virus	7–14 days	highly virulent viral disease
melioidosis	bacterium: *Pseudomonas pseudomallei*	10–14 days	especially prevalent in Southeast Asia
Plague	bacterium: *Yersinia pestis*	2–8 days	transmitted by fleas associated with rodents
Q fever	rickettsia: *Coxiella burnetii*	10–20 days	patients usually recover after severe pneumonia
Rift Valley fever	virus	2–5 days	transmitted by mosquitoes in sub-Saharan Africa
smallpox	variola virus	10–17 days	no therapy available
tularemia	bacterium: *Francisella tularensis*	3–5 days	caused by one of the most infectious known pathogenic bacteria

infectious disease is called an **epidemic** and the study of the occurrence of disease is called **epidemiology**. Most people are familiar with bubonic plague caused by the bacterium *Yersinia pestis* and with smallpox caused by the variola virus. But these are just a few of the causative biological agents of infectious disease. There are a host of causative agents and each one in some respects could be considered a potential biological weapon. Smallpox was used as a biological weapon by the British forces in the French and Indian Wars. Probably the most notorious incident involved Lord Jeffrey Amherst, who gave blankets from smallpox victims to American Indians. The resulting epidemics killed as many as 50% of the members of the afflicted tribes. During World War II, Japanese troops dropped fleas infected with the plague bacterium *Y. pestis* over populations in China.

Potential Biological Weapons

Most likely biological weapons are microorganisms, including algae, fungi, bacteria, and viruses or toxins produced by microorganisms. There are some exceptions. For instance, ricin is a toxin produced from the seed of the castor oil plant. In order to understand their potential as biological weapons we need to understand the general characteristics of each of these groups of microorganisms.

Viruses are in some ways the simplest group of microorganisms, although many might not call viruses microorganisms at all. The argument stems from the fact that viruses can not reproduce on their own; they require other cells in which to reproduce. At minimum, a virus consists of some nucleic acid (DNA or RNA but never both) surrounded by a protein coat. The trick of the virus is to get its nucleic acid into a cell and gain control of the cellular machinery for synthesizing proteins and nucleic acids. It can then use that machinery to synthesize copies of its own nucleic acids and proteins and thereby make new copies of itself. Viruses are responsible for lots of common diseases, including both colds and influenza (the flu). Viruses that have been considered as potential biological weapons include those that cause Congo-Crimean hemorrhagic fever, Ebola hemorrhagic fever, Rift Valley fever, and Venezuelan equine encephalitis. We will consider the potential of the smallpox virus as a potential biological weapon in the next section.

Unlike viruses, bacteria are single-celled organisms capable of growing and dividing (that is, reproducing) entirely on their own. To do so, they need a source of energy and nutrients to provide elements such as carbon, nitrogen, and phosphorus. Their genetic information is stored as DNA but the DNA is not contained within a membrane bound nucleus. The information contained in

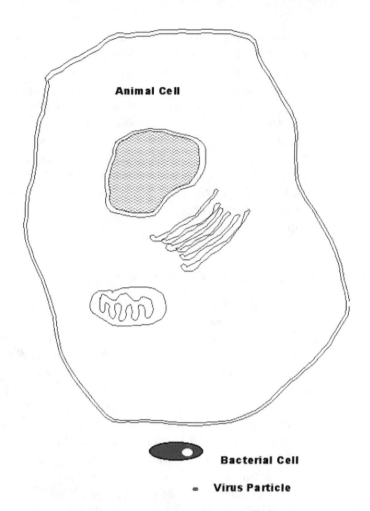

FIGURE 3. Relative sizes of an animal cell, a bacterial cell, and a pox virus. A *B. anthracis* cell is 1.5 to 2 μm long while the pox virus is 0.2 to 0.4 μm long.

the DNA is used to synthesize proteins that function as enzymes that control metabolic processes within the cell. Some bacteria from the genera *Rickettsia* and *Chlamydia* are intracellular parasites and like viruses require other cells to reproduce. Some of the bacteria that have been considered as biological weapons include *Brucella melitensis* (brucellosis), *Vibrio cholera* (cholera), *Pseudomonas pseudomallei* (melioidosis), *Yersinia pestis* (plague), *Coxiella burnetii* (Q fever), and *Francisella tularensis* (tularemia). We will consider the potential of the bacterium *Bacillus anthracis* as a potential biological weapon shortly.

There are far fewer fungi that have been considered potential biological weapons. In part this is because while there are some fungi that are human pathogens they tend to cause diseases such as athlete's foot and ringworm. On the other hand some fungi do produce rather potent toxins. These are called mycotoxins and will be discussed briefly later. In addition, many fungi are plant pathogens and could be used as biological weapons to destroy crops on which humans depend. For example, wheat stem rust is caused by the fungus *Puccinia graminis* and sporadic epidemics have led to losses of up to 200 million bushels of wheat in a single growing season in the United States.

Finally, biological toxins, those chemical compounds that produce disease, can be produced by a wide variety of organisms. A person does not have to become infected with the organism producing the toxin to become sick; exposure to the toxin alone is enough to cause the disease. Such exposure to the toxin is called intoxication. The mode of action of toxins can range from impacting the nervous system (neurotoxins) to impacting the small intestine (enterotoxins). For example, saxitoxin is a neurotoxin produced by marine dinoflagellates. These organisms are responsible for the infamous red tides. Filter-feeding bivalves can accumulate enough of the toxin to threaten the life of people eating the shellfish. This intoxication is known as paralytic shellfish poisoning. Saxitoxin is a thousand times more powerful than the nerve gas sarin and in fact is one of the most potent non-protein toxins yet discovered. Death, when it occurs, is by respiratory paralysis.

Another type of toxin is staphylococcal enterotoxin B, which is produced by the bacterium *Staphylococcus aureus*. Eating food contaminated with this toxin causes distinct gastrointestinal distress but is not usually fatal. The collection of symptoms is commonly referred to as food poisoning. There are also toxins produced by fungi, including the trichothecene mycotoxins. These toxins are potent inhibitors of protein synthesis and other vital cell functions. They are alleged to have been used in Southeast Asia in the "yellow rain" incidents, although whether these incidents represented biological warfare or natural occurrence remains controversial. We will consider the potential of the toxin produced by the bacterium *Clostridium botulinum* as a potential biological weapon later in the text.

The Virus that Causes Smallpox

Prior to an eradication program begun in 1967 by the World Health Organization, smallpox was a global problem. After infection with the smallpox virus, fever is followed by skin eruptions that eventually form pustules. The only known hosts for the smallpox virus are humans. In an unvaccinated population the fatality rate from smallpox can be as high as 30%. Smallpox

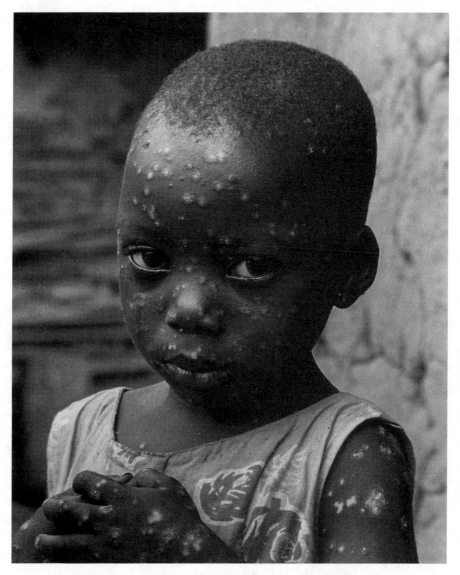

FIGURE 4. A small child displaying the crusty, round pustules that develop from the rash associated with smallpox infection. Source: World Health Organization.

is spread by person-to-person contact, usually as aerosols, and by contact with contaminated surfaces such as bedding or clothing. Transmission of the virus appears to be most rapid once the rash associated with the disease has developed, usually between 7 and 10 days. By this time individuals may already be bedridden with fever and other symptoms. Fortunately, this may reduce the number of contacts with the infected person and slow the spread of the disease.

Only a few viruses are required to produce the infection. This DNA virus is among the largest and most complicated of all viruses. Interestingly the DNA of this virus is replicated in the cytoplasm of the host cell, not in the nucleus. The actual name of the virus that causes smallpox is variola virus (in the genus orthopoxvirus). Other members of this genus include monkeypox, vaccinia, and cowpox viruses. The observation by Edward Jenner in 1796 that milkmaids that had been infected with cowpox (a much milder disease in humans) were resistant to the smallpox virus led to the development of the first vaccine. The actual **vaccine** was developed from the vaccinia virus, hence the name vaccine. *Vaccinus* is Latin for "of cows."

VACCINES

The idea behind a vaccine is to stimulate the body's immune system in a way that can be remembered and used to fight off disease at a later date. We know that the immune system does have a memory because we know that there are diseases that a person does not contract twice. For instance, once someone has had chicken pox and recovered he or she is not susceptible to contract that same disease. We say that they are now immune—that is, exempt or not affected. So an individual who is immune to a particular disease is not affected by the causative agent of that disease. This is as true for the rhinoviruses that cause common colds as it is for the chicken pox virus. Unfortunately there are more than 150 different rhinoviruses so we continue to catch colds years after year, although we do tend to get fewer colds as we age because we are becoming immune to more and more different rhinoviruses.

How does this memory of the immune system work? Well, without going into too much detail, a large part of the answer comes from the white blood cells that make major contributions to keeping us healthy. Our knowledge of how these cells work to help rid the body of infectious agents is expanding rapidly and is contributing to making immunology an exciting and growing field of biology. The white blood cells are able to differentiate into several different cell types with a variety of different functions. One of the major functions is to distinguish between cells that are part of the body—that is, self—and things that are foreign to the body. Things that are foreign would include toxins, viruses, and bacteria. Some white blood cells are responsible for marking foreign particles for elimination. Other white blood cells are responsible for ingesting and destroying bacteria. Still other white blood cells are responsible for recognizing and destroying cells that have been infected by viruses. Finally, some

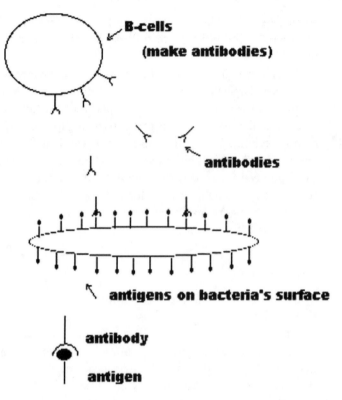

FIGURE 5. Interaction between antibodies and antigens. There is a highly specific fit between antibodies and the antigens they recognize that has been likened to a hand and glove.

white blood cells are responsible for producing **antibodies**, or proteins that are capable of neutralizing or inactivating foreign particles.

One of the things that is most amazing about the immune system is its specificity. Very specific structures on the foreign particles are recognized by the immune system. Those structures that stimulate cells of the immune system to produce antibodies are referred to as **antigens**. The antibodies are able to recognize only the antigen that stimulated their production. Of all the cells in the immune system only a very few recognize the antigens present on any particular infectious agent. Those cells are stimulated to multiply very rapidly. Of course this takes some time, which is why the infectious agent can gain a foothold in the first place and cause the symptoms of the disease. Eventually the immune system gains the upper hand and the infectious agent is eliminated. After this the number of most of the white blood cells returns to normal. But a few of the cells that specifically recognize the foreign infectious agent remain behind; these are called memory cells. When an individual is re-exposed to the same infectious agent it takes less time to mobilize the immune system because memory cells that recognize the foreign agent are already in place. The infectious agent cannot gain a foothold and we say that the individual is now immune to that disease.

Now let's go back to the idea of stimulating the immune system without producing the disease. We might be able to expose individuals to a weakened form of the infectious agent that would be recognized and could reproduce—but would not produce—the symptoms of the disease. Such live but attenuated, or weakened, vaccines are very effective at stimulating immunity but are also the most dangerous because it is possible that the attenuated infectious agent might regain its ability to cause the symptoms of the disease. We could also expose individuals to killed infectious agents. These inactivated (killed) vaccines are safer because the infectious agent is no longer able to reproduce (although there is the possibility that the preparation might contain a few infectious agents that were not killed). The oral Sabin polio vaccine was an attenuated vaccine while the Salk polio vaccine was an inactivated vaccine. Finally, it is possible to expose individuals to only a portion of the infectious agent that contains the antigen in a way that stimulates the immune system. Antigen-based vaccines are the safest but they tend to be less effective.

The vaccinia (smallpox) vaccine currently in use is a live preparation of the vaccinia virus that has been freeze-dried. It does not contain the smallpox virus (variola) at all just the closely related vaccinia virus. Immunity is conferred by cross-reactivity among closely related viruses because neutralizing antibodies produced by vaccinia vaccine are genus-specific and cross-protective against monkeypox, cowpox, and smallpox viruses. The vaccine is administered by taking up a small amount onto a double-pointed needle and repeatedly puncturing the skin.

GLOBAL ERADICATION

It was the effectiveness of the vaccinia vaccine in conferring immunity to smallpox that allowed for the possibility of global eradication. But two other factors were equally important. One was that smallpox does not have any other known host so it could hide in other mammals or insects or anywhere. The other factor was the effectiveness of the quarantine program. As mentioned before, the eradication program was begun in earnest in 1967. The last reported naturally occurring case of smallpox came from Somalia in 1977. There have been sporadic cases of smallpox since that time but they have been the result of laboratory accidents. For a while, stocks of the smallpox virus were kept in at least 19 laboratories in four different countries. Ultimately, the stocks were consolidated in two locations; The Institute of Virus Preparations in Moscow, Russia, and the Centers for Disease Control and Prevention in Atlanta, Georgia. The stocks in Russia are now held in the city of Novosibirsk. There are 461 isolates in the United States' collection and 161 in the Russian lab collection.

Destroying the remaining stocks could conceivably achieve the goal of eradicating the smallpox virus from the planet. Several committees, including those of the World Health Organization, have recommended that the stocks be destroyed. However, arguments have also been made against destroying the stocks:

- If the stocks of the virus are destroyed future research studies of the virus will be impossible;

- The virus has some genes that are similar in sequence to important human genes and without the virus it may be more difficult to understand the function of their gene products.

- Closely related viruses may mutate in ways that would lead to epidemics and understanding the ways in which the smallpox virus causes disease may help us understand these other viruses.

- Destroying the stocks in Russia and the United States does not ensure that the virus does not exist in other laboratories or in nature.

The arguments in favor of destroying the stocks seem a little more straightforward. If the virus escaped from the laboratories, new epidemics could result. Because vaccination programs ended more than twenty-five years ago, it is possible that smallpox epidemics could lead to millions of deaths. It is not known how much immune protection remains for people who were vaccinated twenty-five or more years ago and also a large fraction of the world population was born after smallpox was essentially eradicated and therefore has never been vaccinated.

It is also true that the DNA sequence of the virus is now fairly well known. The viral genome is approximately 200,000 base pairs long. More than 600,000 base pairs of sequence from at least three different strains of the virus have been obtained. In addition, several segments of the genome have been cloned. These cloned fragments by themselves do not pose any safety risk but could be used to learn more about the function of specific gene products.

The CDC is working diligently to use the most modern tools of molecular biology to study the variola virus. The smallpox research is intended to finish mapping the variola genome, to create laboratory diagnostic tests for smallpox, and to find new drugs for treatment if the disease returns. Concerns that the smallpox virus may be used as a biological weapon of mass destruction have intensified the need for research on variola virus diagnostics and clinical interventions have prompted calls for production of additional vaccine. Important questions still remain as to who should be vaccinated and under what circumstances vaccination is warranted. There is strong consensus that vaccinating the general public against smallpox would not be appropriate at this time.

There have been allegations in recent years that stocks of the smallpox virus exist in locations other than the two designated laboratories. It has been suggested that destroying the official stocks might send a strong message to anyone contemplating the use of smallpox as a biological weapon. Ultimately destroying (or not destroying) the smallpox virus stocks is a political decision, but not one that can be made in the absence of a sound understanding of the science involved.

The Bacterium that Causes Anthrax

Anthrax is an infectious disease primarily of hoofed animals, although all mammals are susceptible to varying degrees. The causative agent of anthrax is the bacterium *Bacillus anthracis*. From a microbiological perspective *B. anthracis* has an interesting history. It was the first microorganism to be conclusively shown to be the causative agent of a disease. In 1876, Robert Koch demonstrated that *B. anthracis* caused anthrax in cattle. It was also used to develop the first vaccine based on a bacterium. Louis Pasteur used a live attenuated strain of *B. anthracis* to develop a successful vaccine against anthrax.

B. anthracis is able to produce endospores, which are a highly resistant, dormant form of the bacterium. Bacterial endospores differ from the spores of organisms such as fungi and ferns in several important ways. The spores of bacteria are produced within a single vegetative cell. The spore contains a complete copy of the bacterial genome and all of the machinery necessary to produce a new vegetative cell. The process of a cell producing a spore is called **sporulation** and the process of a spore becoming a new vegetative cell is called **germination**. In between these two processes, spores can survive in the environment for very long periods of time. We already know that *B. anthracis* spores survived for at least 36 years on Gruinard Island. There are claims by researchers in the scientific literature that spores of other *Bacillus* species have survived in ice crystals for 250 million years. These claims are controversial but they do suggest that bacterial spores are long lived.

In addition, endospores are very resistant to heat, dessication, ultraviolet light, and even some harsh chemicals. One reason for this resistance is that

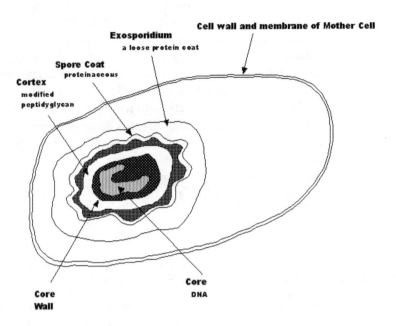

FIGURE 6. Schematic diagram of a bacterial endospore. The multiple layers and low water content both contribute to its unusual resistance.

the spores have very low water content. Much of the water is replaced by calcium and a compound called dipicolinic acid. The lower water content means that heat is transferred less readily and also that chemical reactions that require water are much less likely to occur. The spore is also surrounded by several protective coats or layers. Some of these are made up of peptidoglycan, which is the same material that makes up the cell wall of vegetative cells, and some of these are made up of proteins. The endospores of *B. anthracis* are small ovals between 1.0 and 1.5 µm in length. The spores have a tendency to clump because of electrostatic interactions on their surface and are quite hydrophobic.

It is quite clear that the spores of *B. anthracis* can survive in soil for long periods of time. It is less clear whether vegetative cells of *B. anthracis* are active in the soil. However, when spores are ingested by various mammals the spores are able to germinate and the resulting vegetative cells are able to reproduce, leading to infection. The vegetative cells produce several substances that contribute to the development of disease. Spores that reach the soil, either through defecation or death of the animal, are then available to resume the cycle.

PATHOGENESIS OF *B. ANTHRACIS*

The **pathogenicity** of an organism is its ability to cause disease. We will focus here on the mechanisms that *B. anthracis* employs to cause disease in humans. First there are three basic forms of the disease in humans: cutaneous, gastrointestinal, and inhalational. In each case, spores need to germinate to form vegetative cells. The signals for germination are an environment rich in amino acids and nitrogenous bases, such as the purines and pyrimidines. These conditions are well met within the human body. In the case of inhalational anthrax the spores can actually germinate with macrophages. This is a nasty trick because macrophages are a type of white blood cell whose job is to remove infectious agents. Here we have spores germinating in the macrophage and then being carried to lymph nodes where they can lead to systemic disease.

An important question is how many spores are required to cause infection? The figure that you most often see quoted is that the LD_{50} is between 8,000 and 10,000 spores. LD_{50} stands for **Lethal Dose 50** and is the number of spores required to kill 50% of the test animals. The estimates for LD_{50} come from studies of occupational exposure primarily among wool sorters and goat-hair mill workers. Additional information for the estimates has also been gleaned from accidental exposures, such as the one in Sverdlovsk, and from the deliberate exposure of non-human mammals, such as occurred on Gruinard Island. Keep in mind that these are statistical estimates of the dose that might be expected to lead to 50% mortality; they do not represent a minimum threshold below which disease cannot occur. In specific cases exposures below 8,000 spores could well lead to development of anthrax and in other cases exposures above 10,000 spores may not lead to the development of anthrax.

After germination, *B. anthracis* produces two different toxins. One toxin leads to edema, or swelling, and the other leads to necrosis, or cell death. The genes that code for the toxins are carried on a plasmid. **Plasmids** are small circular pieces of DNA that are found in many bacteria. The plasmid that codes for the toxins in *B. anthracis* is called pX01. Three separate genes code for three proteins that work together to make the two toxins. The proteins are called protective antigen, edema factor, and lethal factor. All three are proteins with molecular weights between 80,000 and 90,000 daltons. Protective antigen binds to the surface of the cell and facilitates the entry of itself and both edema factor and lethal factor into the cell. Once inside, the edema factor acts as an adenyl cyclase, an enzyme that catalyzes the conversion of ATP into cyclic-AMP. Cyclic-AMP is an important signaling molecule in the cell and its overproduction leads to water imbalance and swelling. Lethal factor acts as a protease, an enzyme that cleaves other proteins. In this case the protein that is cleaved is another enzyme, MAP kinase. Kinases are enzymes that add phosphate to

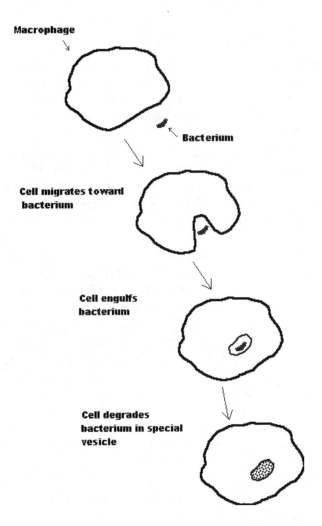

FIGURE 7. A macrophage engulfing a bacterial cell. *B. anthracis* spores can survive this process and can even germinate with the macrophage.

other proteins and MAP kinase is so important in regulating cellular processes that its loss can lead to cell death.

There is a second plasmid that is important to the pathogenicity of *B. anthracis*. This second plasmid (pX02) doesn't code for toxins but it does code for a capsule that surrounds the bacterium. The capsule, which is composed of polyglutamic acid, helps to protect the bacterium from elimination by the immune system. Polyglutamic acid is a polymer in which the amino acid glutamate is the only repeating unit. Interestingly, the bonds linking the glutamic acid repeating units are different than the bonds linking amino acids in proteins. It is possible for plasmids to be lost and bacteria that lose pX02 can no longer produce a capsule and are no longer pathogenic.

As many as 95% of the naturally occurring cases of anthrax are cutaneous anthrax. Spores usually enter through a break in the skin, very often after contact with contaminated wool or goat hides as previously discussed. The initial symptom is a papule, or small pimple, that develops on the surface of the skin. The papule enlarges into an ulcer by the second day and ultimately develops into a painless black depression called an eschar. The black eschar gives *B. anthracis* its name: *anthrakis* is the Greek word for coal. As many as 2,000 cases of cutaneous anthrax occur each year worldwide. Without antibiotic therapy as many as 20% of cutaneous anthrax cases would be fatal.

Gastrointestinal anthrax may develop from spores germinating in either the upper GI tract or the lower GI tract. The former case leads to the formation of ulcers in the nose, mouth, or throat, and extensive swelling. The latter can lead to nausea, vomiting, and bloody diarrhea. Gastrointestinal anthrax is most often caused by eating uncooked (or undercooked) contaminated meat. This form of anthrax is relatively uncommon, although outbreaks have occurred in Asia and Africa. Mortality from gastrointestinal anthrax may be as high as 50%.

The incubation period for inhalation anthrax is thought to be 1 to 6 days, but the germination of spores that have been carried to the lymph nodes may be delayed by up to 60 days. This may explain why there were new cases being reported up to 43 days after the spore release in the Sverdlovsk incident. Once the vegetative cells begin to reproduce and release the toxins, hemorrhaging, swelling, and necrosis follow. The toxins are so powerful that they can lead to death even after all the bacteria have been cleared from the system by antibiotics.

ANTIBIOTICS AND VACCINES

B. anthracis is susceptible to several common antibiotics. **Antibiotics** are substances produced by microorganisms that have an inhibitory effect on other microorganisms. We did not mention antibiotics in our discussion of smallpox because there are no antibiotics that are effective against viruses. This is also why your physician is not likely to prescribe antibiotics if you are suffering from either a cold or the flu (both are caused by viruses). A good antibiotic for treating anthrax is doxycycline, which is in a class of antibiotics known as the tetracyclines. Tetracyclines are inhibitors of protein synthesis. Fortunately they inhibit protein synthesis in prokaryotic organisms such as bacteria, but not in eukaryotic organisms such as humans. They are, however, broad-spectrum antibiotics, meaning that they are effective against a wide range of bacteria.

Bacteria are able to develop resistance to antibiotics through repeated exposure. Luckily, *B. anthracis* has not been exposed to antibiotics too frequently and therefore has not acquired much resistance. Bacterial resistance to a particular antibiotic may involve preventing the antibiotic from getting into the cell, changing the target of the antibiotic, or acquiring the ability to pump the antibiotic back out of the cell. Bacteria are even able to acquire enzymes that can degrade antibiotics. The prevalence of antibiotic resistance through overuse or misuse of antibiotics is increasingly becoming a problem. Therefore, health care professionals are reluctant to use broad spectrum antibiotics if there are other antibiotics that are equally effective against a particular bacterium. That is why the antibiotic of choice for treating anthrax has been ciprofloxacin (cipro). Ciprofloxacin is a member of the class of fluoroquinolones that interfere with enzymes of DNA replication. The course for antibiotic treatment after suspected exposure to *B. anthracis* is 60 days—because of the problem of delayed spore germination previously discussed. Until they germinate, spores are dormant and for the most part antibiotics are effective only against growing cells.

We have already mentioned that Pasteur developed a vaccine against anthrax using an attenuated strain of *B. anthracis*. In the 1940s, a new live attenuated vaccine (based on the Sterne strain) was developed for veterinary use. Vaccines based on live attenuated strains are considered too dangerous for use in humans throughout most of the world, although they may get sporadic use in some countries. The Sterne strain of *B. anthracis* was used by the Aum Shinrikyo cult in the deliberate release mentioned earlier.

The vaccine in current use in the United States is made from a strain that can no longer produce a capsule and therefore is non-pathogenic. After growth in liquid culture the cells are filtered out. It is believed that protective antigen that has been released from the cells and is now in solution is responsible for conferring immunity. Remember an antigen is the portion of the infectious agent that stimulates the immune system. Protective antigen is a part of the toxin of *B. anthracis* but it is only toxic in association with edema factor and lethal factor. The vaccine has to be administered in six separate doses over an eighteen-month period.

The Toxin of Clostridum Botulinum

Botulism, the illness caused by *Clostridium botulinum*, is an intoxication that is often associated with canned and foil-wrapped foods. This is because *C. botulinum* is an anaerobic bacterium and can grow only in the absence of oxygen.

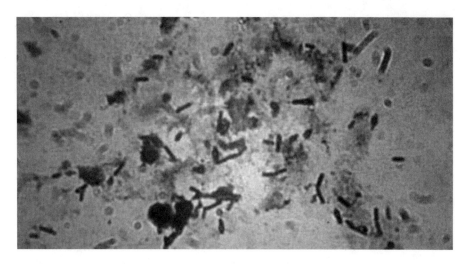

FIGURE 8. Photomicrograph of *C. botulinum*. The cells have been stained to increase their contrast.

Therefore, situations in which air is excluded give the bacterium a chance to grow and produce the toxin. This bacterium also produces endospores and commonly can be found in soils. The toxin is the most powerful poison known. At least seven different structural forms of the toxin have been identified and they have been assigned the letters A through G. The strain of *C. botulinum* that is most common in the United States produces toxin A.

The toxin consists of two protein chains, the larger of which binds to the surface of the motor neurons and helps move the smaller of the two chains across the membrane and into the cell. The smaller of the two chains can act as an endopeptidase and can cleave other proteins once it gets into the cell. The net result is that within nerve cells the vesicles that contain the neurotransmitter acetylcholine cannot fuse with the cell membrane properly and signals can therefore not be propagated from one nerve cell to the next. This produces a flaccid muscle paralysis. The binding of the toxin is irreversible and people who recover from botulism do so because their cells regenerate new connections.

Because the toxin is a protein, it is easily destroyed by heat. However, because the organism produces spores, special care needs to be taken to ensure that foods are treated in a way that will kill any spores present. Otherwise, if conditions are right the spores may germinate and new vegetative cells will produce toxin that could lead to intoxication if the food is not reheated prior to consumption. Antibiotics cannot be used to treat intoxication because the toxin has already been released from the bacterium. In addition to foodborne botulinum intoxication, *C. botulinum* can also grow in wounds and in the human intestine, two situations in which reduced oxygen concentrations can occur. These two infectious diseases can be treated with antibiotics because the bacterium is actively growing in these situations. Intentionally aerosolized botulinum toxin has been shown to cause an inhalation form of botulism in primates.

Death from botulism is usually caused by obstructed airways and failure of the respiratory muscles. Therefore, primary therapy must involve supporting the function of the impacted systems. Supportive actions can include tube or intravenous feeding and mechanical ventilation. There are also antitoxins available. Antitoxins are antibodies that were produced in another organism using the toxin as the antigen. In the case of botulinum toxin, the toxin is usually injected into a horse. The toxin is modified to make it less toxic and in this modified form is called a toxoid. The neutralizing antibodies, which the horse makes in response to the toxoid, can be purified and used as a therapy. Each of the seven different forms of the toxin produce antibodies that are specific to that form, so often the antitoxin that is given as a therapy contains antibodies to more than one form. The antitoxin has no effect on toxin that has already bound (and therefore can't reverse paralysis that has already occurred), but antitoxin can prevent any free toxin from causing additional damage. Because of this, botulinum antitoxin may be more useful as a preexposure preventative than as a postexposure therapy.

Botulinum toxin has also been licensed in the United States for several therapeutic uses. These therapies rely on the ability of the toxin to cause localized flaccid muscle paralysis. The licensed therapies include torticollis (wryneck), strabismus (lazy eye), and blepharospasm (abnormal blinking or eyelid tic). Botulinum toxin also is used off label for ailments such as migraine headaches, lower back pain, and cerebral palsy. (Off label refers to uses of an approved drug for conditions for which it was not approved.) The botulinum toxin has just recently been approved for cosmetic use. An advertisement on the Internet reads:

> Whenever we frown, we gather tissue between our eyebrows into a fold. For many of us, this causes a chronic furrow producing a frustrated and angry look on our face. This can be alienating to others, and upsetting to ourselves. Now, you can make these frown lines disappear without surgery and without scars, by a simple treatment of Botox. Botox, the trade name of Botulinum Toxin Type A, originally was used for the treatment of strabismus (lazy eye) and blepharospasm (uncontrolled eye blinking). Now it is being used by plastic surgeons to achieve a younger appearance. It accomplishes this by blocking the nerve impulses to muscles, which then results in a

form of temporary muscle paralysis. Our physicians inject an extremely small dosage of Botox directly into the muscle. The action of that specific muscle is then stopped, ridding the area of problem frown lines or twitches. (http://internetmd.com/botox.html)

What the advertisement does not say is that you would be injecting into your forehead the most poisonous substance known. Of course the therapeutic doses are 0.3% of the estimated lethal inhalational dose and 0.005% of the estimated lethal oral dose.

The Epidemiology of Biological Weapons

We began by saying that bioweapons should be unthinkable. But there can be no denying that military powers including the United States and the former Soviet Union have studied the use of biological weapons. There also can be no denying that several smaller nations have access to microorganisms and to toxins that could be used as biological weapons. And finally there can be no denying that there have been isolated attempts to use biological weapons as agents of terror.

It has been argued that only by studying biological weapons can we protect ourselves. To the extent that this is true, the biological factors that need to be considered are very much the same as those of naturally occurring diseases. The primary factors are:

- infectivity
- pathogenicity
- virulence
- toxicity
- transmissibility
- incubation period

Infectivity is the ability of a microorganism to grow on or within a host. It is important to distinguish between infection and disease. All humans are infected with a variety of microorganisms, many of which do not cause us any harm at all; in fact many of these infections are beneficial to us. There are also many microorganisms infecting (living within or on) us that are capable of causing disease but are normally held in check by the body's defenses. These are the opportunistic pathogens that take advantage of us when our defenses are compromised. Even organisms that are frank pathogens can differ in the ability to infect humans.

This is why we also need to consider **pathogenicity** or the ability to cause disease. For bacteria and fungi the two major mechanisms of pathogenesis are invasion of tissue and production of toxins. For viruses the major mechanisms are destruction of cells and interference with physiological functions. Virulence is a measure of how likely infection with a particular organism will lead to death of the host. It is usually measured as LD_{50} as previously discussed. Toxicity is the damage caused to a host by a toxin. Some toxins are released from active cells, while other toxins are released only when the cells die and lyse (disintegrate).

Transmissibility and incubative period have a big impact on how quickly and how widely a disease spreads among a population. From the perspective of the pathogen there are three parts to transmissibility: escape from the current host, movement to a new host, and infection of the new host. Some diseases can be transferred by person-to-person contact. Sexually transmitted diseases fall into this category, as do many respiratory diseases that are transmitted as fine droplets. Pathogens may also be transmitted indirectly, carried by vectors such as insects or rodents. We generally call diseases that are transmitted by direct or indirect contact contagious diseases. The incubation period is the amount of time between the initial exposure and the appearance of symptoms. This plays an important role in the spread of disease. Asymptomatic individuals who are contagious are more like to contact other individuals and spread the disease.

There are additional considerations when we shift our thinking from the natural occurrence of disease to the use of biological weapons. For example, spores of *B. anthracis* can be found in soils but they rarely cause disease in humans. Recently there have been discussions of the concept of weaponizing *B. anthracis* spores. The spores tend to clump together and in clumps of 100 or more are much less likely to cause anthrax. Also, in order to expose large numbers of people to the possibility of inhaling sufficient numbers of spore it is necessary to develop a way to get the spores to stay suspended in the air as aerosols. Viruses, on the other hand, are often transmitted as aerosol (remember the common cold virus) but they tend not to persist for long periods of time outside of the cells in which they replicate. This would complicate any attempt to expose large numbers of people to a pathogenic virus. The botulinum toxin is very potent, but it is easily broken down by heat or standard water treatment procedures. These are just a few of the types of considerations that would be involved in the use of microorganisms as biological weapons.

And while it is true that understanding biological weapons may be the best way to defend against such weapons, the argument has also been made that studying biological weapons makes their use more likely. For instance, an effective vaccine against a particular pathogen could be used to immunize individuals intending to deploy the pathogen as a weapon. In nature there is always a war going on. It is a war between organisms that are pathogens and their hosts. Pathogens evolve to take advantage of their hosts in an effort to increase their own numbers. In response the hosts evolve better defenses, and the pathogen then counterattacks with other measures. But in the end it is a war that the pathogen cannot win, for if the pathogen is too successful it will have no host and it depends on the host. Humans, as the host to many pathogens, have been participating in this war for a long time. Efforts to improve hygiene and the development of antibiotics are two good examples. So while we are active participants in the war between pathogens and host, there is no good reason for us to intervene on the side of the pathogens through the development or use of biological weapons.

Randall W. Phillis
Department of Biology
University of Massachusetts
Amherst, MA 01003
E-mail: rphillis@bio.umass.edu

Steve Goodwin
Department of Microbiology
639 North Pleasant St.
University of Massachusetts
Amherst, MA 01003-5720
E-mail: sgoodwin@microbio.umass.edu

Resources about Biological Weapons

ESPECIALLY FOR STUDENTS

American Society for Microbiology (http://www.asmusa.org/pcsrc/bioprep.htm) The premier scientific organization in the world for the study of microorganisms presents a site devoted to resources related to biological weapons control and bioterrorism preparedness.

Center for Civilian Biodefense (http://www.hopkins-biodefense.org/index.html) A center of The Johns Hopkins University sponsored by The Alfred P. Sloan Foundation & The Robert Wood Johnson Foundation.

Centers for Disease Control and Prevention (http://www.cdc.gov/) A wealth of information, including public health emergency response and preparedness plans.

Morbidity and Mortality Weekly (http://www.cdc.gov/mmwr/) The Centers for Disease Control and Prevention's weekly publication. This site provides an opportunity to see the practice of epidemiology in action.

Recent Research in the Journal *Nature* (http://www.nature.com/nature/anthrax/) Anthrax and *Bacillus anthracis* are the subject of much ongoing research and several reports have recently been published in *Nature,* one of the world's premier scientific journals.

RESOURCES FOR EDUCATORS

Academic Info (http://www.academicinfo.net/terrorismbio.html) Academic Info is a user-supported, nonprofit organization that provides internet access to educational resources. They maintain a web page on chemical and biological terrorism and weapons.

Center for the Study of Traumatic Stress Uniformed Services University, Disaster Care Resources (http://www.usuhs.mil/psy/disasterresources.html) This site, maintained by Public Health Services, provides resources for coping with the stresses of biological terrorism.

Agents of Bioterror (http://www.pbs.org/wgbh/nova/bioterror/agents.html) This Public Broadcasting System site examines eight potential biological weapons and has a special section for teachers.

Working Group on Civilian Biodefense (http://www.ama-assn.org/ama/pub/category/6232.html) In 1999, the Working Group on Civilian Biodefense developed some consensus-based recommendations for measures to be taken by medical and public health professionals if biological weapons were used against a civilian population. Separate reports were prepared dealing with anthrax, tularemia, smallpox, botulinum toxin, and plague. These are available online through the *Journal of the American Medical Association* website.

REVIEW ARTICLES

Bhatnagar, R. and S. Batra. 2001. Anthrax toxin. *Crit. Rev. Microbiol.* 27:167–200. A review focused on the mode of action of the anthrax toxin.

Meselson, M., J. Guillemin, H.-J. Martin, A. Langmuir, I. Popova, A. Shelkov, and O. Yampolskaya. 1994. *Science.* 266:1202–1208. The story of the incident in Sverdlovsk as investigated and told by a preeminent biologist.

Mock, M. and A. Fouet. 2001. Anthrax. *Ann. Rev. Microbiol.* 55:647–671. A comprehensive review of our current scientific understanding of *B. anthracis* and anthrax.

Young, J. A. and R. J. Collier. 2002. Attacking anthrax. *Scientific American* 186:48–59. A discussion of using the latest discoveries (including the sequence of the *B. anthracis* genome) to fight anthrax. (http://www.sciam.com/2002/0302issue/0302young.html)

POPULAR BOOKS

Alibek, K. and S. Handelman. *Biohazard: The Chilling True Story of the Largest Covert Biological Weapons Program in the World—Told from Inside by the Man Who Ran It.* Random House Inc., 1999.

Ewald, P. W. *Plague Time.* Simon & Schuster, 2000.

Miller, J., S. Engelberg, and W. J. Broad. *Germs: Biological Weapons and America's Secret War.* Simon & Schuster, 2001.

Neese, R. M. and G. C. Williams. *Why We Get Sick.* Random House, 1994.

Tucker J. B. *Scourge, The Once and Future Threat of Smallpox.* Atlantic Monthly Press, 2001.